Introducti...

Math
Grade 5

This book has been designed to help all students succeed in math. It is part of a *Basics First* series that provides students with opportunities to practice basic math skills that will help them begin to understand many of the mathematical properties that they will use over and over throughout their lives. The activities can be used as supplemental material to reinforce any existing math program.

The activities in this book have been created to help students feel confident in math computation. In order to ensure success, students must be guided through the activities presented. Most of the activities feature an amusing illustration which will motivate students and help maintain a high level of interest as students complete them. Allow students to use models at any time while working on the activities in this book.

A variety of fun and simple formats are included throughout the book. Students will enjoy solving cross number puzzles, decoding messages, solving riddles, and much more.

The skills covered in this book can be taught in the classroom or at home. They include place value through billions and thousandths; rounding, multiplying, and dividing decimals; adding, subtracting, multiplying, and dividing whole numbers and fractions; circumference of circles; area of parallelograms; volume of cubes and rectangular prisms; and many more.

Rules to Remember

Divisibility rules

Name _____

Use the divisibility rules below to complete this page.

> 2 = If the last digit is an even number, the number is divisible by 2.
> 3 = If the sum of the digits can be divided by 3, the number is divisible by 3.
> 5 = If the last digit is 0 or 5, the number is divisible by 5.
> 9 = If the sum of the digits can be divided by 9, the number is divisible by 9.
> 10 = If the last digit is 0, the number is divisible by 10.

Draw a (cloud) around the numbers that are divisible by 2.

Draw a (heart) around the numbers that are divisible by 3.

Underline the numbers that are divisible by 5.

Draw a (star) around the numbers that are divisible by 9.

(Circle) the numbers that are divisible by 10.

(Hint: Some numbers will be divisible by more than one number.)

1. 270	6. 940	11. 1,485	16. 2,106
2. 815	7. 8,136	12. 3,010	17. 1,825
3. 562	8. 2,012	13. 616	18. 442
4. 914	9. 735	14. 500	19. 9,003
5. 539	10. 1,044	15. 1,500	20. 386

© Frank Schaffer Publications, Inc.

FS-30105 Math

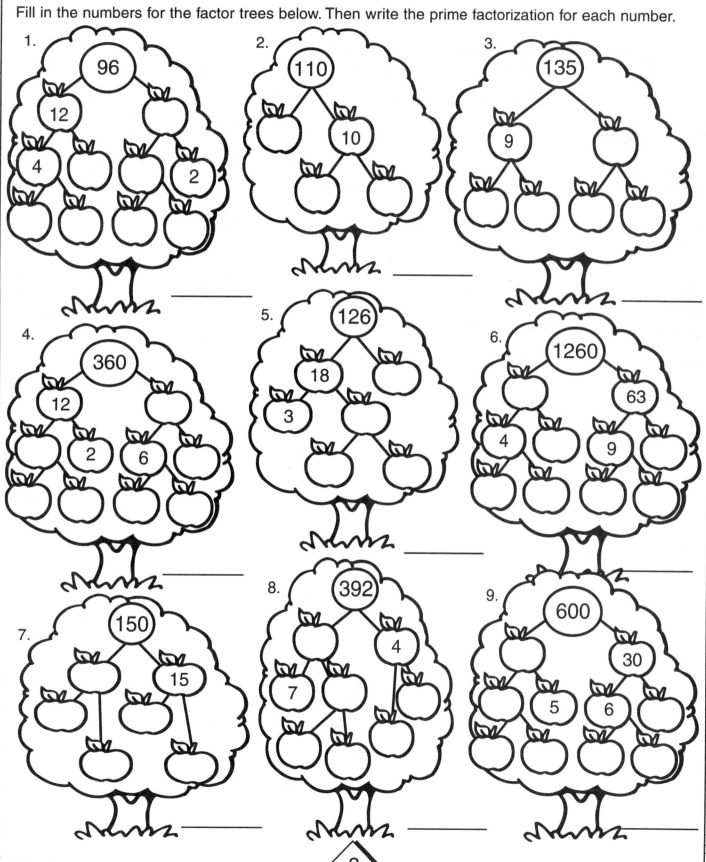

Musical Mathematics

Place value through billions

Name _____

What did the opera singer have in his mouth?

To answer the riddle, write the word name for each number. Then read the circled letters from top to bottom.

478,620,341 = ○___ ___ ___   ___ ___ ___ ___ ___ ___   ___ ___ ___ ___ ___ ___ ___ -
___ ___ ___ ___ ___ ___ ___   ___ ○___ ___ ___ ___ ___ ,   ___ ___ ___   ___ ___ ___ ___ ___ ___ ___ ___ -
___ ___ ___ ___   ___ ___ ___ .

2,163,075,918 = ___ ___ ___   ___ ○___ ___ ___ ___ ___ ___ ,   ___ ___ ○___ ___ ___ ___ ___ ___   ___ ___ ___ ___ ___ ___ ___ -
○___ ___ ___ ___ ___ ___ ,   ___ ___ ___ ___ ○___ ___ ___ ___   ___ ___ ___ ___ ___ ___ ___ ___ .

8,022, 202, 022 = ○___ ___ ___ ___ ___   ___ ___ ___ ___ ___ ___ ,   ___ ___ ___   ___ ___ ___ ___ ___ ___ -
___ ___ ___   ___ ___ ___ ___ ___ ,   ___ ○___ ___ ___ ___ ___ ___ ___ -

6,504,011,900 = ___ ___ ○___ ___ ___ ___   ___ ___ ___ ___ ___ ___ ___ ,   ___ ___ ___ ___   ○___ ___ ___ ___ ___ ___ ,
___ ___ ___ ___ ___ ___   ___ ___ ___ ___ ___ ___ ___ ___ .

3,120,471,032 = ___ ___ ___ ___ ___   ___ ___ ___ ___ ___ ___ ,   ___ ___ ___   ___ ○___ ___ ___ ___ ___ ___ ___
___ ___ ___ ___ ___ ___   ___ ___ ___ ___ ___ ___ ___ ___ - ___ ___ ○___   ___ ___ ___ .

A Whale of a Puzzle

Place value through thousandths

Name _____

Write the standard form of each of the following to complete the puzzle. (Hint: Decimals take up their own box.)

Down

1. five and four hundred six thousandths
2. eleven and three hundred fifty-six thousandths
4. three hundred seventy-one and one hundred three thousandths
7. one hundred eighteen and seven hundred four thousandths
8. seven and ninety-six thousandths
10. sixty-two and nine hundred three thousandths
12. fifty-seven thousandths

Across

2. one and eight hundred twelve thousandths
3. seven hundred thirteen thousandths
5. seventy-four and fifty-nine thousandths
6. eighty-one and seven hundred seventy-eight thousandths
9. five hundred sixteen and three hundred thirty-eight thousandths
11. nine thousandths
13. eighty-eight and thirteen thousandths

© Frank Schaffer Publications, Inc.

FS-30105 Math

Racy Rounding

Rounding decimals

Name _____

What is the speed limit in Egypt?

To answer the riddle, round each decimal to the nearest thousandth. Then write the corresponding letter above the correct decimal at the bottom of the page.

X. 21.007624   Y. 21.471028   I. 21.80458

U. 21.09531   E. 21.9983   S. 21.4657

A. 21.04458   O. 21.007201   T. 21.00183

N. 21.982385   L. 21.10453   R. 21.9989

S. 21.898989   N. 21.8037   I. 21.4648   H. 21.000973

___ ___ ___ ___ ___   ___ ___ ___ ___ ___
21.899 21.805 21.008 21.002 21.471   21.982 21.465 21.105 21.998 21.466

          ___ ___   ___ ___ ___ ___
          21.045 21.804   21.001 21.007 21.095 21.999

© Frank Schaffer Publications, Inc.   6   FS-30105 Math

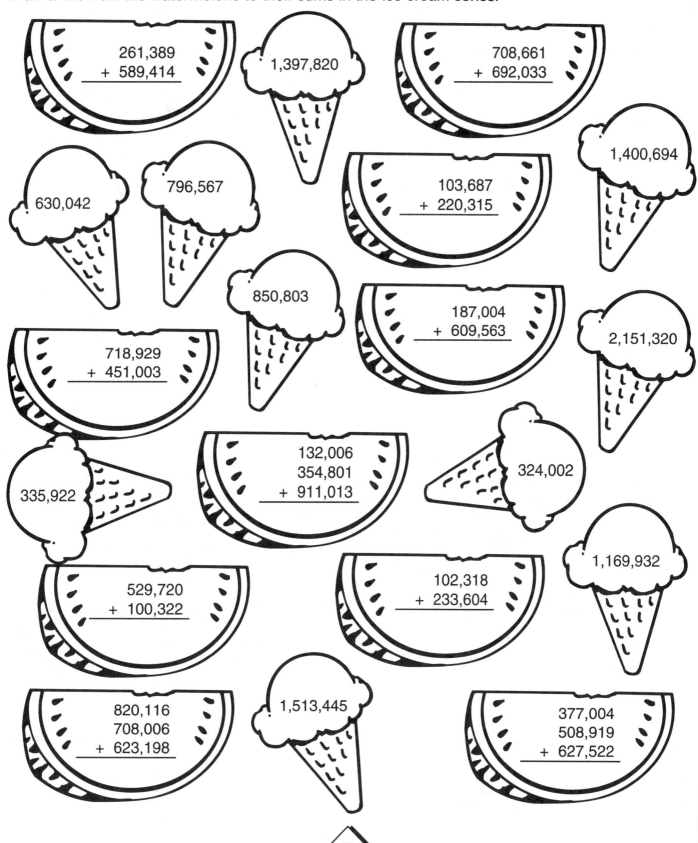

Loony Legislation

Subtracting 4- to 6-digit numbers

Name _____

There are many strange laws that exist all over the world. To find out something that is illegal in Detroit, Michigan, find the differences below at the bottom of the page. Put the corresponding letter above each.

H. 618,720 − 59,606	L. 129,011 − 100,973	O. 54,188 − 13,909	O. 438,715 − 9,886

H. 733,842 − 523,051	N. 629,704 − 7,890	N. 211,333 − 198,466	A. 97,800 − 14,555

E. 386,512 − 317,109	E. 500,999 − 2,891	T. 317,433 − 205,617	S. 46,781 − 18,665

G. 438,002 − 199,311	I. 503,726 − 122,399	Z. 298,772 − 289,697	W. 710,020 − 687,304

B. 128,777 − 93,605	I. 310,821 − 299,097	T. 165,712 − 100,888	O. 721,078 − 520,019

111,816   40,279   _____   28,116   12,887   201,059   428,829   9,075   69,403

22,716   210,791   11,724   28,038   498,108

35,172   83,245   64,824   559,114   381,327   621,814   238,691

© Frank Schaffer Publications, Inc.   8   FS-30105 Math

More Loony Legislation

Greatest common factors

Name _____

Another wacky law is in Leahy, Washington. To find out what it is, find the greatest common factors of the pairs of numbers below. Use the "Decoder Box" to spell out the law.

24, 22   36, 42   30, 14   54, 60   30, 55   40, 32

____   ____   ____   ____   ____   ____

24, 60   24, 27   18, 78   70, 30   44, 36   10, 18   21, 35

____   ____   ____   ____   ____   ____   ____

16, 20   26, 44   26, 65   55, 88   36, 84   30, 21   28, 24   99, 36

____   ____   ____   ____   ____   ____   ____   ____

52, 39   33, 36   98, 56   45, 54   56, 48   65, 35

____   ____   ____   ____   ____   ____

Decoder Box

2 = N   6 = O   11 = U
3 = L   7 = G   12 = B
4 = I   8 = E   13 = P
5 = S   9 = C   14 = A
           10 = W

Java Joe

Multiplying by 3-digit numbers

Name _____

What kind of coffee does Joe the undertaker drink?

To answer the riddle, find the products below at the bottom of the page. Put the corresponding letter above each.

F. 3,819
  x   506

A. 298
  x 187

O. 562
  x 419

T. 4,008
  x   825

E. 21,003
  x   561

D. 30,006
  x   571

E. 870
  x 611

C. 728
  x 841

D. 972
  x 316

N. 2,109
  x   919

F. 718
  x 422

I. 673
  x 228

___ ___ ___ ___ ___ ___ ___ - ___
17,133,426 11,782,683 612,248 235,478 302,996

___ ___ ___ ___ ___ ___ ___
1,932,414 153,444 1,938,171 55,726 3,306,600 531,570 307,152

Super Scientific

Dividing by 2-digit numbers

Name _____

What did the scientist get when she crossed an oven and a mattress?

To answer the riddle, find the quotients below at the bottom of the page. Put the corresponding letter above each.

K. 13)6,004   N. 22)11,198

R. 38)8,800   F. 17)2,132   E. 26)14,014   E. 43)17,802

A. 11)3,011   I. 62)20,708   D. 52)32,136   S. 44)39,732

B. 14)10,064   A. 21)3,806   B. 33)9,867   T. 55)21,326

| 299 | 231R22 | 414 | 181R5 | 461R11 | 125R7 | 273R8 | 903 | 387R41 |

| 334 | 509 | | 718R12 | 539 | 618 |

© Frank Schaffer Publications, Inc.   11   FS-30105 Math

Mathematical Meteorology

Adding mixed numbers without renaming

Name _____

What was the weather report in Mexico?

To find the answer to the riddle, add the mixed numbers below. Write the corresponding letter above the sum at the bottom of the page.

A. $4\frac{1}{8} + 5\frac{1}{3} =$ _____   A. $5\frac{1}{6} + 4\frac{1}{4} =$ _____

I. $7\frac{2}{9} + 3\frac{1}{10} =$ _____   A. $6\frac{2}{5} + 5\frac{1}{4} =$ _____

L. $6\frac{1}{8} + 6\frac{1}{6} =$ _____   I. $2\frac{3}{10} + 10\frac{1}{5} =$ _____   O. $4\frac{4}{15} + 7\frac{2}{3} =$ _____

Y. $6\frac{3}{20} + 4\frac{1}{4} =$ _____   M. $9\frac{1}{4} + 2\frac{2}{7} =$ _____   L. $7\frac{2}{9} + 1\frac{5}{12} =$ _____

H. $2\frac{5}{12} + 6\frac{1}{3} =$ _____   T. $5\frac{2}{5} + 5\frac{1}{6} =$ _____   C. $6\frac{3}{10} + 3\frac{1}{2} =$ _____

T. $9\frac{11}{20} + 3\frac{1}{4} =$ _____   E. $1\frac{1}{3} + 10\frac{2}{15} =$ _____   T. $5\frac{5}{16} + 3\frac{1}{4} =$ _____

O. $1\frac{3}{7} + 8\frac{1}{5} =$ _____   D. $3\frac{5}{12} + 7\frac{1}{4} =$ _____   H. $4\frac{3}{7} + 4\frac{1}{2} =$ _____

$\overline{9\frac{4}{5}}$ $\overline{8\frac{13}{14}}$ $\overline{10\frac{29}{90}}$ $\overline{8\frac{23}{36}}$ $\overline{12\frac{1}{2}}$ $\overline{10\frac{17}{30}}$ $\overline{11\frac{14}{15}}$ $\overline{10\frac{2}{3}}$ $\overline{9\frac{11}{24}}$ $\overline{10\frac{2}{5}}$,

$\overline{8\frac{3}{4}}$ $\overline{9\frac{22}{35}}$ $\overline{12\frac{4}{5}}$ $\overline{8\frac{9}{16}}$ $\overline{9\frac{5}{12}}$ $\overline{11\frac{15}{28}}$ $\overline{11\frac{13}{20}}$ $\overline{12\frac{7}{24}}$ $\overline{11\frac{7}{15}}$

© Frank Schaffer Publications, Inc.

Swiss Sentences

Adding mixed numbers with renaming

Name _____

Complete the cheesy number sentences below.

$7\frac{4}{5}$	+	$2\frac{3}{8}$	=			$2\frac{1}{2}$
		+				+
$5\frac{5}{6}$	+	$3\frac{3}{4}$	=			$2\frac{9}{10}$
+		=		+		=
$3\frac{2}{5}$				$1\frac{2}{3}$	+		=
=				=		
		$4\frac{13}{16}$	+		=	
		+				+
$6\frac{1}{2}$	+	$5\frac{5}{8}$	=			$1\frac{47}{48}$
+		=		+		=
$10\frac{19}{20}$				$7\frac{11}{12}$		
=				=		
		$1\frac{59}{60}$	+		=	

Kitty Cat Cuisine

Dividing whole numbers by fractions

Name _____

To answer the following riddles, divide the whole numbers by fractions. Put the corresponding letter above the quotient at the bottom of each set of problems.

A. What do cats love to have for breakfast?

S. $6 \div \frac{5}{8}$ E. $12 \div \frac{3}{4}$ C. $5 \div \frac{5}{7}$ I. $10 \div \frac{4}{9}$

R. $8 \div \frac{2}{3}$ S. $9 \div \frac{6}{11}$ P. $3 \div \frac{2}{7}$ M. $6 \div \frac{4}{5}$

I. $7 \div \frac{1}{3}$ I. $4 \div \frac{8}{9}$ E. $5 \div \frac{3}{5}$ C. $8 \div \frac{4}{9}$

$\overline{7\frac{1}{2}}$ $\overline{22\frac{1}{2}}$ $\overline{7}$ $\overline{8\frac{1}{3}}$ $\overline{18}$ $\overline{12}$ $\overline{4\frac{1}{2}}$ $\overline{9\frac{3}{5}}$ $\overline{10\frac{1}{2}}$ $\overline{21}$ $\overline{16}$ $\overline{16\frac{1}{2}}$

B. What does a cat call a bird that he's eaten for breakfast?

E. $15 \div \frac{3}{4}$ E. $2 \div \frac{6}{7}$ E. $15 \div \frac{5}{6}$

H. $9 \div \frac{1}{6}$ D. $14 \div \frac{7}{8}$ W. $10 \div \frac{2}{5}$

T. $9 \div \frac{4}{5}$ D. $7 \div \frac{3}{5}$ R. $20 \div \frac{4}{11}$

T. $10 \div \frac{1}{12}$ D. $5 \div \frac{5}{6}$ S. $6 \div \frac{2}{9}$ E. $8 \div \frac{3}{8}$

$\overline{27}$ $\overline{54}$ $\overline{55}$ $\overline{2\frac{1}{3}}$ $\overline{11\frac{2}{3}}$ $\overline{16}$ $\overline{18}$ $\overline{6}$ $\overline{120}$ $\overline{25}$ $\overline{21\frac{1}{3}}$ $\overline{20}$ $\overline{11\frac{1}{4}}$

© Frank Schaffer Publications, Inc.   14   FS-30105 Math

Troublesome Tulips

Least common denominator

Name _____

Find the least common denominator and write the like fractions in the leaves.

A. 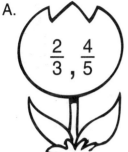 $\frac{2}{3}, \frac{4}{5}$ $\frac{2}{3}, \frac{1}{6}$ 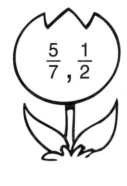 $\frac{5}{7}, \frac{1}{2}$  $\frac{1}{5}, \frac{5}{6}$  $\frac{5}{8}, \frac{1}{3}$

B.  $\frac{3}{4}, \frac{2}{5}$ 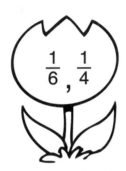 $\frac{1}{6}, \frac{1}{4}$ 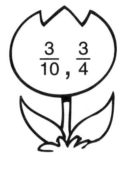 $\frac{3}{10}, \frac{3}{4}$  $\frac{5}{12}, \frac{1}{6}$ 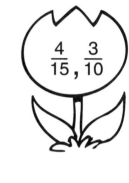 $\frac{4}{15}, \frac{3}{10}$

C. 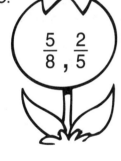 $\frac{5}{8}, \frac{2}{5}$ 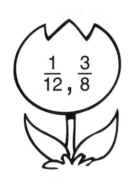 $\frac{1}{12}, \frac{3}{8}$ 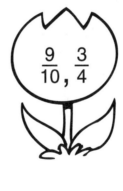 $\frac{9}{10}, \frac{3}{4}$ 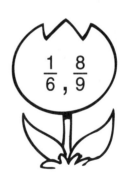 $\frac{1}{6}, \frac{8}{9}$  $\frac{13}{15}, \frac{5}{6}$

D. 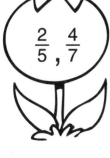 $\frac{2}{5}, \frac{4}{7}$ 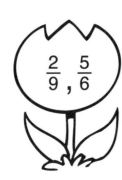 $\frac{2}{9}, \frac{5}{6}$ 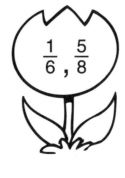 $\frac{1}{6}, \frac{5}{8}$ 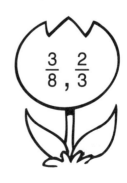 $\frac{3}{8}, \frac{2}{3}$  $\frac{7}{25}, \frac{7}{10}$

Diving for Decimals

Fractions and percents

Name _____

Draw lines from the fractions in the masks to the equivalent percents in the fins.

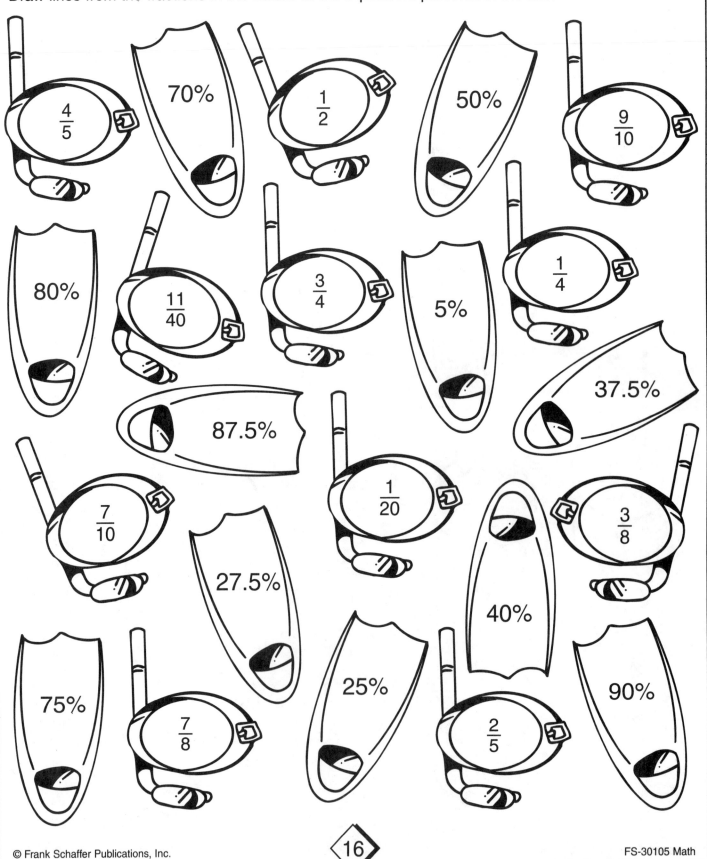

Creepy Crawlies

Writing mixed numbers as fractions

Name _____

Write the mixed numbers below as fractions.

- $2\frac{7}{10}$
- $5\frac{4}{5}$
- $8\frac{2}{3}$
- $6\frac{3}{7}$
- $4\frac{8}{9}$
- $4\frac{2}{11}$
- $20\frac{1}{2}$
- $7\frac{5}{8}$
- $3\frac{1}{5}$
- $15\frac{1}{3}$
- $11\frac{7}{8}$
- $10\frac{3}{4}$
- $12\frac{1}{9}$
- $2\frac{7}{15}$
- $6\frac{5}{6}$

© Frank Schaffer Publications, Inc.

FS-30105 Math

Monkey Business

Multiplying whole numbers and fractions

Name _____

Draw lines from the chimpanzees to the correct products in the bananas.

Fragrant Fractions

Multiplying fractions

Name _____

What do skunks read?

To answer the riddle, multiply the fractions below. Put the corresponding letter above the product at the bottom of the page. (Hint: Some letters will not be used.)

E. $\frac{3}{8} \times \frac{2}{3}$ N. $\frac{4}{5} \times \frac{15}{16}$ S. $\frac{5}{6} \times \frac{4}{15}$ E. $\frac{2}{3} \times \frac{5}{6}$

J. $\frac{3}{8} \times \frac{2}{9}$ L. $\frac{3}{4} \times \frac{2}{5}$ R. $\frac{5}{12} \times \frac{4}{15}$ I. $\frac{5}{13} \times \frac{2}{20}$

T. $\frac{5}{8} \times \frac{2}{3}$ S. $\frac{9}{10} \times \frac{2}{3}$ B. $\frac{5}{6} \times \frac{3}{20}$ E. $\frac{4}{5} \times \frac{5}{6}$

A. $\frac{7}{12} \times \frac{6}{8}$ M. $\frac{2}{5} \times \frac{1}{4}$

S. $\frac{4}{9} \times \frac{3}{4}$ L. $\frac{2}{5} \times \frac{3}{8}$

$\frac{1}{8}$ $\frac{5}{9}$ $\frac{2}{9}$ $\frac{5}{12}$ $\quad\underline{}\quad$ $\frac{3}{5}$ $\frac{1}{10}$ $\frac{2}{3}$ $\frac{3}{20}$ $\frac{3}{10}$ $\frac{1}{4}$ $\frac{1}{9}$ $\frac{1}{3}$

Pizza for Polly

Multiplying mixed numbers

Name _____

Polly is wild about pizza. To find out what she has on her pizza, multiply the mixed numbers by the pizza toppings below. She ordered the ingredients whose products are correct. Write the correct answers to the problems that are wrong.

pepperoni $\quad 2\frac{1}{2} \times 3\frac{2}{5} = 8\frac{1}{2}$

black olives $\quad 2\frac{1}{4} \times 5\frac{1}{3} = 12$

bacon $\quad 1\frac{3}{8} \times 5\frac{1}{3} = 7\frac{2}{3}$

mushrooms $\quad 1\frac{1}{6} \times 3\frac{3}{4} = 4\frac{3}{8}$

sausage $\quad 5\frac{1}{4} \times 1\frac{2}{7} = 7\frac{1}{4}$

pineapple $\quad 2\frac{3}{10} \times 1\frac{3}{7} = 3$

green peppers $\quad 2\frac{1}{10} \times 8\frac{1}{3} = 17\frac{1}{2}$

onions $\quad 3\frac{1}{8} \times 2\frac{2}{5} = 7\frac{1}{2}$

tomatoes $\quad 1\frac{1}{20} \times 5\frac{5}{7} = 7$

Canadian bacon $\quad 2\frac{4}{15} \times 22\frac{1}{2} = 48$

anchovies $\quad 8\frac{2}{3} \times 4\frac{1}{2} = 39$

jalapeño peppers $\quad 2\frac{5}{6} \times 1\frac{1}{2} = 4\frac{1}{4}$

extra cheese $\quad 4\frac{2}{3} \times 1\frac{1}{5} = 5\frac{2}{5}$

meatballs $\quad 1\frac{5}{6} \times 2\frac{2}{5} = 4\frac{2}{5}$

List the ingredients Polly had on her pizza.

_____

_____

And More Loony Legislation

Subtracting mixed numbers without renaming

Name _____

It is illegal to eat something in public on Sunday in Kansas. To find out what it is, find the differences below and put the letter above the answer at the bottom of the page.

N. $14\frac{5}{6} - 8\frac{1}{2}$ L. $9\frac{3}{4} - 2\frac{3}{20}$ K. $8\frac{5}{6} - 2\frac{2}{15}$

E. $11\frac{1}{2} - 3\frac{3}{10}$ A. $5\frac{8}{9} - 1\frac{2}{3}$ T. $20\frac{4}{5} - 14\frac{4}{15}$

T. $16\frac{11}{12} - 12\frac{3}{8}$ E. $11\frac{2}{3} - 4\frac{7}{15}$ E. $10\frac{9}{10} - 5\frac{3}{4}$

A. $20\frac{2}{5} - 13\frac{1}{6}$ A. $21\frac{3}{4} - 17\frac{2}{3}$ R. $19\frac{5}{8} - 14\frac{1}{3}$

M. $11\frac{1}{2} - 6\frac{1}{3}$ T. $20\frac{7}{9} - 12\frac{1}{2}$ S. $18\frac{5}{6} - 10\frac{4}{9}$

$\overline{5\frac{7}{24}}$ $\overline{4\frac{2}{9}}$ $\overline{4\frac{13}{24}}$ $\overline{8\frac{5}{18}}$ $\overline{7\frac{3}{5}}$ $\overline{7\frac{1}{5}}$ $\overline{8\frac{7}{18}}$ $\overline{6\frac{1}{3}}$ $\overline{7\frac{7}{30}}$ $\overline{6\frac{7}{10}}$ $\overline{5\frac{3}{20}}$

$\overline{5\frac{1}{6}}$ $\overline{8\frac{1}{5}}$ $\overline{4\frac{1}{12}}$ $\overline{6\frac{8}{15}}$

Sailing Subtraction

Subtracting mixed numbers from whole numbers

Name _____

Which sailboat is going to win the race? To find out, find the differences in the sailboats. The sailboat with the largest number as the difference wins. Color this boat as festively as you'd like.

- $26 - 8\frac{4}{5}$
- $21 - 2\frac{4}{11}$
- $39 - 19\frac{2}{9}$
- $26 - 6\frac{14}{25}$
- $32 - 13\frac{5}{8}$
- $26 - 7\frac{3}{7}$
- $49 - 31\frac{1}{6}$
- $21 - 3\frac{11}{20}$
- $35 - 15\frac{6}{13}$
- $33 - 14\frac{19}{22}$
- $41 - 23\frac{1}{8}$
- $25 - 6\frac{13}{15}$
- $28 - 8\frac{11}{16}$

Flower Power

Subtracting mixed numbers with renaming

Name _____

Using colored pencils, markers, or crayons, color the flowers as indicated by their differences.

$6\frac{2}{3}$ — yellow with green stars

$11\frac{3}{5}$ — pink with yellow polka dots

$11\frac{3}{4}$ — blue and white stripes

$12\frac{3}{5}$ — purple

$13\frac{19}{20}$ — red and white stripes

$8\frac{7}{18}$ — blue and green stripes

$6\frac{9}{10}$ — orange

$8\frac{8}{11}$ — purple and green checks

$10\frac{1}{3}$ — blue

$7\frac{5}{16}$ — pink with white swirls

$7\frac{1}{2}$ — yellow with orange polka dots

$12\frac{8}{15}$ — red with purple stars

Sailing Signals

Dividing decimals by whole numbers

Name _____

Sailors use flag signals to communicate. Find the quotients below by using the given information. Give your answers in flag signals and color them the appropriate colors. Remember to draw your decimal points.

R = red   W = white   B = blue   BL = black   Y = yellow

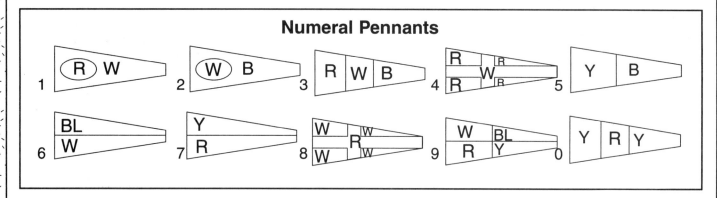

Numeral Pennants

1. [R/W] = 1, [W/B] = 2, [R W B] = 3, [R/R W/R/B] = 4, [Y B] = 5, [BL/W] = 6, [Y/R] = 7, [W/W R W/W] = 8, [W BL/R Y] = 9, [Y R Y] = 0

1. [R/R W/R/B] [R W B] . [W B] ÷ [R W] [W B] =

2. [Y B] . [R W B] [W B] [W/W R W/W] ÷ [W/W R W/W] =

3. [W/W R W/W] [Y R Y] . [Y/R] [R W] ÷ [Y/R] =

4. [Y B] [Y R Y] . [R W B] [R/R W/R/B] ÷ [BL/W] =

5. [W B] [R W B] . [W BL/R Y] [W B] ÷ [W B] [R W B] =

6. [R W] [W BL/R Y] [R W B] [Y R Y] . [R/R W/R/B] ÷ [R/R W/R/B] =

7. [BL/W] [Y/R] . [W B] [Y/R] ÷ [R W B] [R W] =

Caveman Cam

Multiplying decimals by whole numbers

Name _____

Cam, the unfrozen caveman mathematician, left the cave drawings below. Use the box to decipher the multiplication problems.

Key:
- drop = 1
- star = 2
- bone = 3
- lightning = 4
- spiral = 5
- flower = 6
- diamond = 7
- bull = 8
- square = 9
- heart = 0
- circle = ×
- tulip = =

1. $3.6 \times 44 =$
2. $2.15 \times 18 =$
3. $0.68 \times 123 =$
4. $12.9 \times 78 =$
5. $8.54 \times 26 =$
6. $5.7 \times 88 =$
7. $10.1 \times 5 =$
8. $2.99 \times 58 =$
9. $0.76 \times 83 =$
10. $4.38 \times 16 =$

Another Loony Legislation

Multiplying decimals with zeros in the product

Name _____

In Lubbock, Texas, it is illegal to sleep someplace. Find out where by solving the multiplication problems below. Put the corresponding letter above the product at the bottom of the page.

A. 0.156 × 0.2

A. 0.43 × 0.08

A. 0.316 × 0.3

G. 0.07 × 0.02

A. 0.019 × 0.4

E. 0.003 × 5.8

R. 0.011 × 0.06

I. 0.132 × 0.3

N. 0.008 × 0.5

N. 2.22 × 0.001

C. 0.08 × 0.02

B. 0.13 × 0.14

G. 1.22 × 0.04

___ ___   ___
0.0396 0.00222   0.0948

___ ___ ___ ___ ___ ___ ___
0.0488 0.0312 0.00066 0.0182 0.0076 0.0014 0.0174

___ ___ ___
0.0016 0.0344 0.004

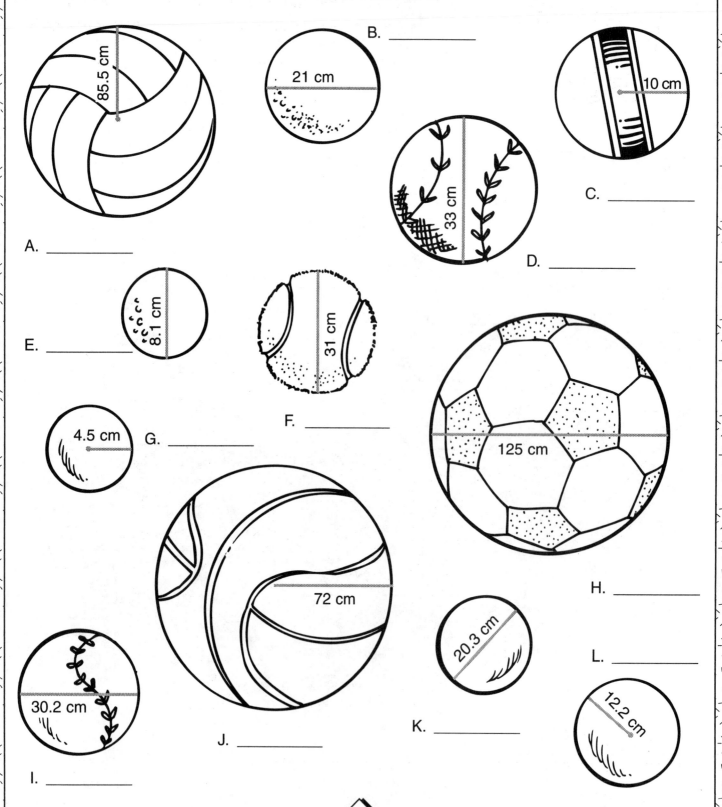

Have a Ball!

Circumference of circles

Name _____

George the Giant loves to play sports. Of course, his equipment is quite large. Find the circumference of the balls below that he uses. Use 3.14 for π.

A. _____
B. _____
C. _____
D. _____
E. _____
F. _____
G. _____
H. _____
I. _____
J. _____
K. _____
L. _____

© Frank Schaffer Publications, Inc.

28

FS-30105 Math

Weird Words

Area of parallelograms

Name _____

What is it called when someone is covered in freckles?

To answer the riddle, find the area of each parallelogram below. Put the letters in each blank at the bottom of the page going from the smallest to the largest area.

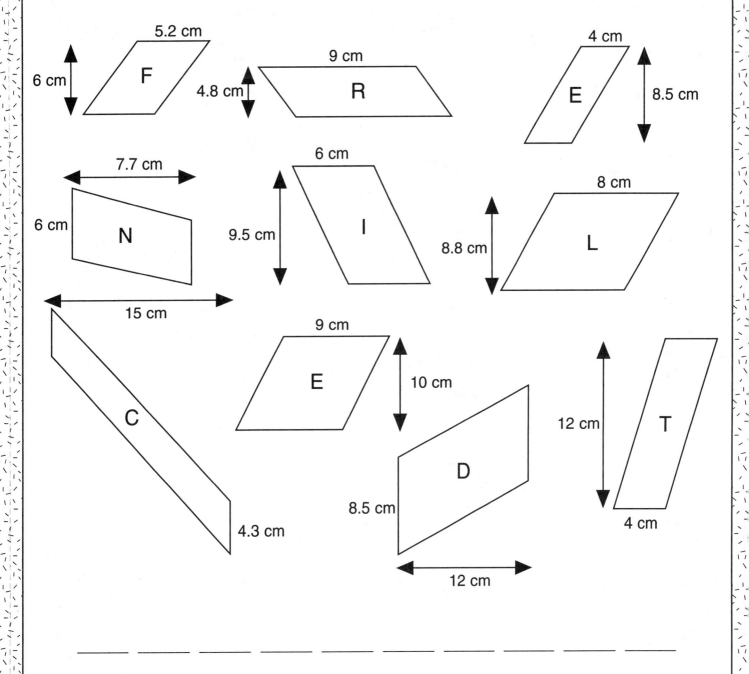

29

FS-30105 Math

More Weird Words

Volume of cubes and rectangular prisms

Name _____

What is the morbid fear of slime?

To answer the riddle, find the volumes below at the bottom of the page. Put the corresponding letter above each.

I. 3 cm × 3 cm × 3 cm

O. 5 cm × 3 cm × 2 cm

N. 4 cm × 2 cm × 2 cm

A. 3 cm × 2 cm × 8 cm

H. 8 cm × 5 cm × 2 cm

O. 4 cm × 4 cm × 4 cm

B. 1 cm × 1 cm × 1 cm

B. 2 cm × 3 cm × 7 cm

P. 3 cm × 3 cm × 4 cm

L. 2 cm × 5 cm × 4 cm

E. 2 cm × 3 cm × 3 cm

N. 7 cm × 2 cm × 4 cm

| 1 cm³ | 40 cm³ | 18 cm³ | 16 cm³ | 56 cm³ | 64 cm³ | 36 cm³ | 80 cm³ | 30 cm³ | 42 cm³ | 27 cm³ | 48 cm³ |

© Frank Schaffer Publications, Inc.

30

FS-30105 Math

Answers

Page 2

The numbers are divisible by:
1. 2, 3, 5, 9, 10
2. 5
3. 2
4. 2
5. none
6. 2, 5, 10
7. 2, 3, 9
8. 2
9. 3, 5
10. 2, 3, 9
11. 2, 3, 9
12. 2, 5, 10
13. 2
14. 2, 5, 10
15. 3, 5, 9
16. 2, 3, 9
17. 5
18. 2
19. 3
20. 2

Page 3

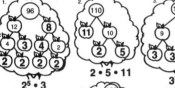

Page 4

478,620,341 = four hundred seventy-eight million, six hundred twenty thousand, three hundred forty-one

2,163,075,918 = two billion, one hundred sixty-three million, seventy-five thousand, nine hundred eighteen

8,022,202,022 = eight billion, twenty-two million, two hundred two thousand, twenty-two

6,504,011,900 = six billion, five hundred four million, eleven thousand, nine hundred

3,120,471,032 = three billion, one hundred twenty million, four hundred seventy-one thousand, thirty-two

Riddle = falsetto teeth

Page 5

Page 6

SIXTY NILES AN HOUR

Page 7

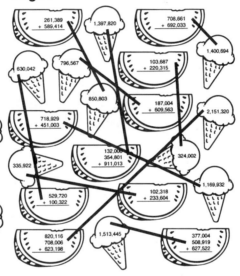

Page 8

TO SNOOZE WHILE BATHING

Page 9

NO NOSE BLOWING IN PUBLIC PLACES

Page 10

DECOFFINATED

Page 11

BREAKFAST IN BED

Page 12

CHILI TODAY, HOT TAMALE

Page 13

$7\frac{4}{5}$	+	$2\frac{3}{8}$	=	$10\frac{7}{40}$		$2\frac{1}{2}$
+				+		+
$5\frac{5}{6}$	+	$3\frac{3}{4}$	=	$9\frac{7}{12}$		$2\frac{9}{10}$
+		=		+		=
$3\frac{2}{5}$		$6\frac{1}{8}$		$1\frac{2}{3}$	+	$5\frac{2}{5}$	=	$7\frac{1}{15}$
=		=		=		
$9\frac{7}{30}$		$4\frac{13}{16}$	+	$11\frac{1}{4}$	=	$16\frac{1}{16}$
		+				+
$6\frac{1}{2}$	+	$5\frac{5}{8}$	=	$12\frac{1}{8}$		$1\frac{47}{48}$
+		=		+		=
$10\frac{19}{20}$		$10\frac{7}{16}$		$7\frac{11}{12}$		$18\frac{1}{24}$
=				=		
$17\frac{9}{20}$		$1\frac{59}{60}$	+	$20\frac{1}{24}$	=	$22\frac{1}{40}$

Page 14

A. MICE CRISPIES
B. SHREDDED TWEET

Page 15

A. 10/15, 12/15; 4/6, 1/6; 10/14, 7/14; 6/30, 25/30; 15/24, 8/24

B. 15/20, 8/20; 2/12, 3/12; 6/20, 15/20; 5/12, 2/12; 8/30, 9/30

C. 25/40, 16/40; 2/24, 9/24; 18/20, 15/20; 3/18, 16/18; 26/30, 25/30

D. 14/35, 20/35; 4/18, 15/18; 4/24, 15/24; 9/24, 16/24; 14/50, 35/50

Page 16

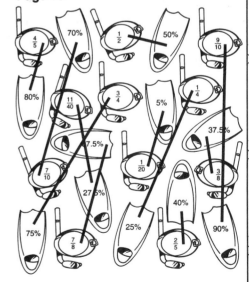

Answers

Page 17

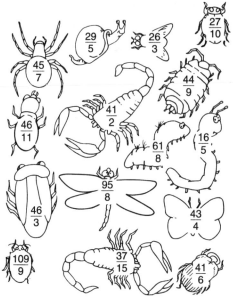

anchovies = correct
jalapeño peppers = correct
extra cheese = 5 3/5
meatballs = correct
Ingredients: pepperoni, mushrooms, black olives, green peppers, onions, anchovies, jalapeño peppers, meatballs

Page 21
RATTLESNAKE MEAT

Page 22

Page 23

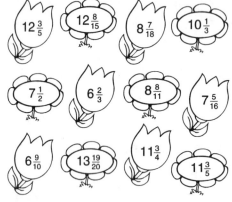

Page 24
1. 43.2 ÷ 12 = 3.6
2. 5.328 ÷ 8 = 0.666
3. 80.71 ÷ 7 = 11.53
4. 50.34 ÷ 6 = 8.39
5. 23.92 ÷ 23 = 1.04
6. 1930.4 ÷ 4 = 482.6
7. 67.27 ÷ 31 = 2.17

Page 25
1. 3.6 x 44 = 158.4
2. 2.15 x 18 = 38.7
3. 0.68 x 123 = 83.64
4. 12.9 x 78 = 1006.2
5. 8.54 x 26 = 222.04
6. 5.7 x 88 = 501.6
7. 10.1 x 5 = 50.5
8. 2.99 x 58 = 173.42
9. 0.76 x 83 = 63.08
10. 4.38 x 16 = 70.08

Page 26

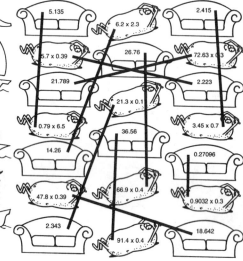

Page 27
IN A GARBAGE CAN

Page 28
A. 536.94 cm
B. 65.94 cm
C. 62.8 cm
D. 103.62 cm
E. 25.434 cm
F. 99.34 cm
G. 28.26 cm
H. 392.5 cm
I. 94.828 cm
J. 452.16 cm
K. 63.742 cm
L. 76.616 cm

Page 29
FERNTICLED

Page 30
BLENNOPHOBIA

Page 18

Page 19
BEST-SMELLERS

Page 20
pepperoni = correct
black olives = correct
bacon = 7 1/3
mushrooms = correct
sausage = 6 3/4
pineapple = 3 2/7
green peppers = correct
onions = correct
tomatoes = 6
Canadian bacon = 51